plaids

A visual survey of pattern variations

Tina Skinner

Schiffer Publishing Ltd

4880 Lower Valley Rd. Atglen, PA 19310 USA

Dedication

To Craig, for lost weekends and infinite understanding.

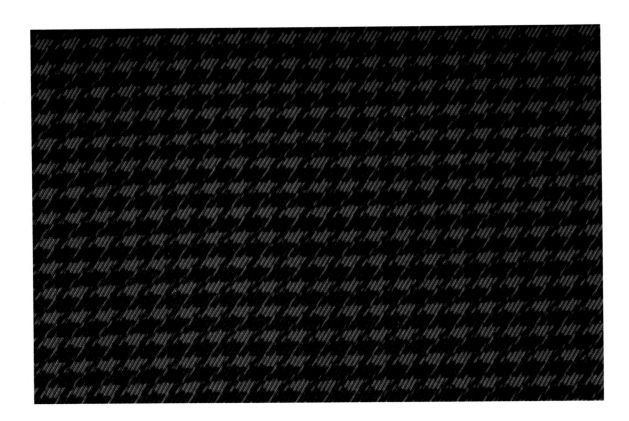

Designed by Bonnie M. Hensley

ISBN: 0-7643-0481-X
Printed in Hong Kong

Published by Schiffer Publishing Ltd.
4880 Lower Valley Road
Atglen, PA 19310
Phone: (610) 593-1777; Fax: (610) 593-2002
E-mail: schifferbk@aol.com
Please write for a free catalog.
This book may be purchased from the publisher.
Please include $3.95 for shipping.

Try your bookstore first.

We are interested in hearing from authors
with book ideas on related subjects.

contents

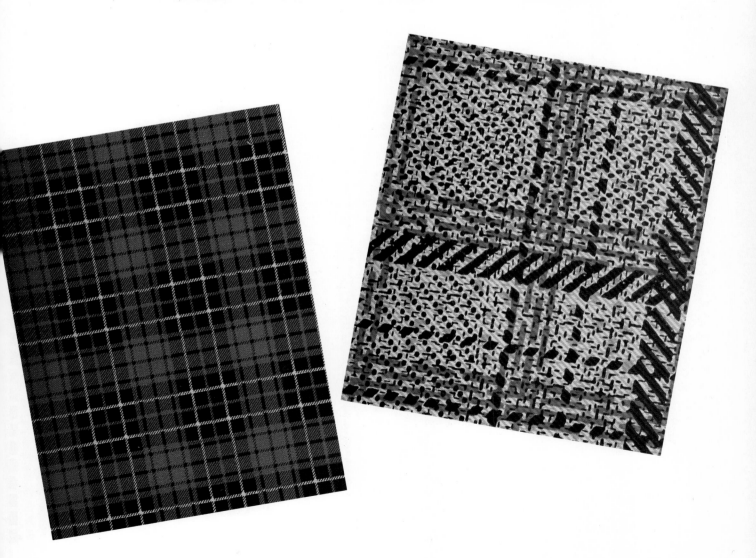

introduction

Once the loom was created, once the weaver had more than one color of yarn in her hands, plaids were inevitable. Simply crossed vertical and horizontal lines have provided infinite fodder for artists' imaginations throughout the history of fabrics. Printing technology multiplied these possibilities even further, spawning an entire genre of fabric design which has been a staple for clothing and household designers for decades.

This book exploited a vast library of fabric swatches documenting more than five decades of fabric design, from simple cotton prints in the early part of the century to the rich colors and design variations that grew along with technology and the post-World War II economies in the United States and Europe.

The book begins with woven fabrics, demonstrating how plaids are a natural outgrowth of the weaving process, before moving on to the way artists have imitated that woven look in more recent prints.

Basic plaid designs are explored, from "windowpane" plaids, formed by straight, intersecting lines, to the dotted effect created by regularly spaced, thicker lines in "gingham" fabrics. The stacking of "squares" is depicted, as are the "check" patterns formed by irregularly spaced, intersecting lines.

Harlequin or "diamond" patterns, formed by gling plaids or stretching them, have been popular for decades among print-cloth designers. "Hound's-tooth" plaid, a distinctive, four-point gingham pattern, got its start back in the misty history of Scotland, but has gone through countless variations in the hands of designers. The rich diversity of designs created by those designers has been illustrated extensively in the large portion of the book devoted to novelty plaids, which fully explores experimentation in printed plaids over the last five decades. In addition, the use of floral patterns with plaid motifs is examined.

Finally, because no book on plaids would be complete without them, an entire section on woven wool Scottish tartans rounds out this extensive collection. In fact, the word "plaid" owes its existence to the Scots, who coined it to describe not the fabric but an article of clothing made from plaid patterns particular to the wearer's family.

For these tartans I am deeply indepted to Highland Heritage, a Scottish specialty store in Wilmington, Delaware, for the loan of fabric swatches to be photographed and for their expertise in the subject.

Though the tartans are all woven wool, the other fabric swatches pictured have varied histories. Each is dated, either by year or decade—whichever was known—and the fabric's country of origin is named. When known, fabric content is also listed. Some are simply labeled as a "blend" because the content was not clear. Since these photographs are of historical fabric swatches, many do not encompass the entire design. It is up to the reader to complete the pattern in his or her own mind's eye. Naturally, it is hoped that this book will spur imaginations and lead to another five decades of creative plaid design.

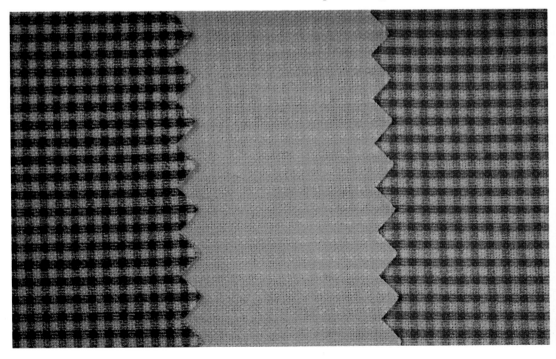

woven and imitation weaves

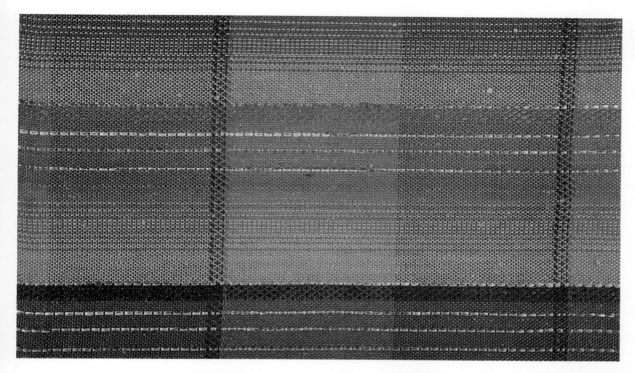

Rayon/polyester.
USA. 1980s.

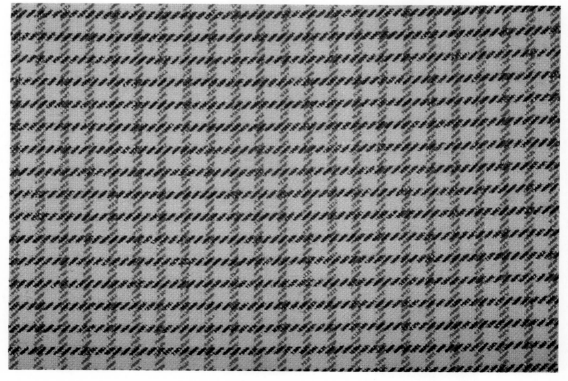

Cotton. USA. 1954.

Woven

Cotton/polyester. USA. 1960s.

Cotton/polyester blend. USA. 1960s.

Cotton/polyester blend. USA. 1960s.

Cotton/polyester blend. USA. 1960s.

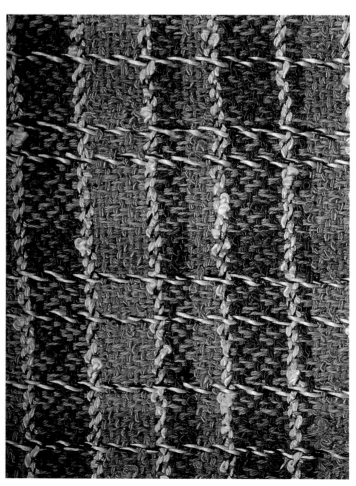

Cotton/polyester blend. USA. 1960s.

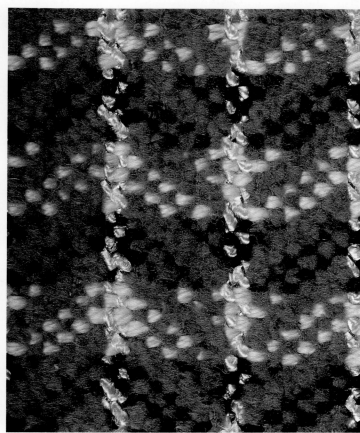

Cotton/polyester blend. USA. 1960s.

Polyester blend. USA. 1960s.

Polyester blend. USA. 1960s.

Polyester blend. USA. 1960s.

Cotton/polyester. USA. 1980s.

Polyester blend. USA. 1960s.

Cotton blend. USA. 1960s.

Dacron/cotton/rayon. USA. 1980s.

Rayon/polyester/cotton. USA. 1980s.

Rayon/polyester/cotton/olefin. USA. 1980s.

Rayon/polyester/cotton/olefin. USA. 1980s.

9

Imitation Weaves

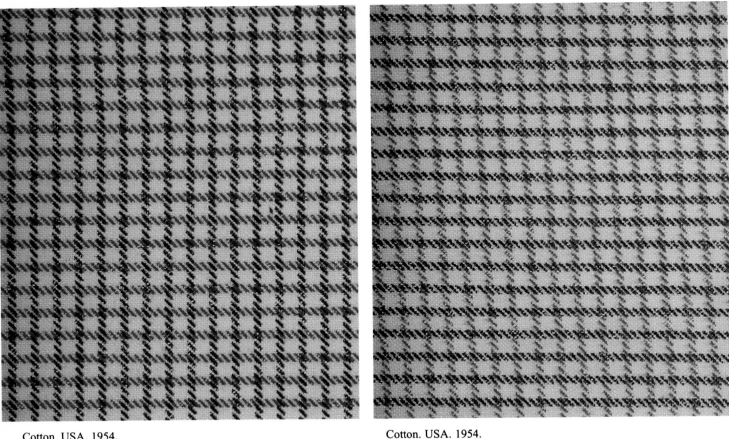

Cotton. USA. 1954.
Cotton. USA. 1954.

Cotton. USA. 1954.
Cotton. USA. 1954.

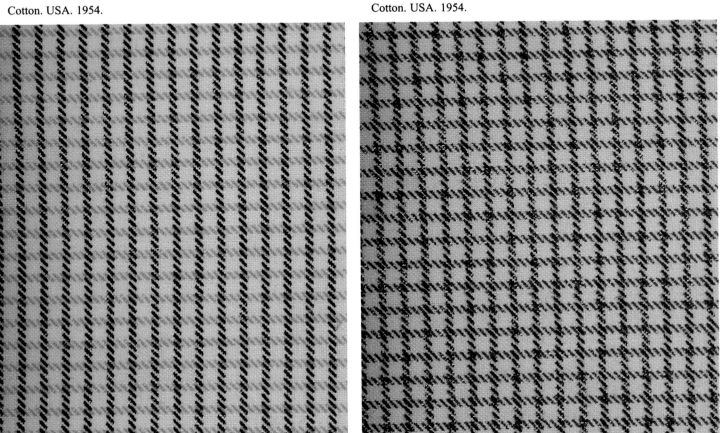

Cotton. USA. 1948.

Cotton. USA. 1948.

Rayon/polyester. USA. 1980s.

Acetate. France. 1967.

Polyester. France. 1966.

Silk/wool. France. 1966.

Polyester/cotton. France. 1964.

Polyester. France. 1966.

Silk/rayon. France. 1966.

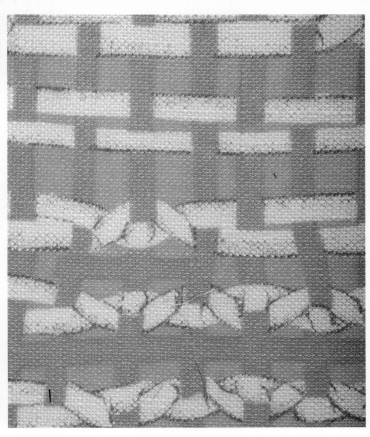

Cotton/polyester. France. 1966.

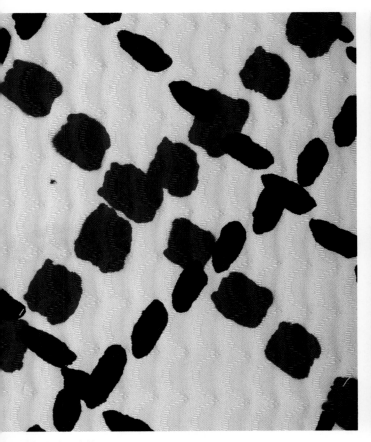

Silk. Italy. 1963.

Silk. France. 1964.

windowpane

Cotton. USA. 1950s.

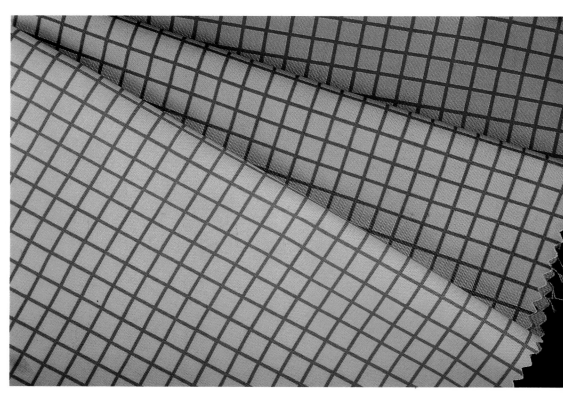

Cotton. USA. 1950s.

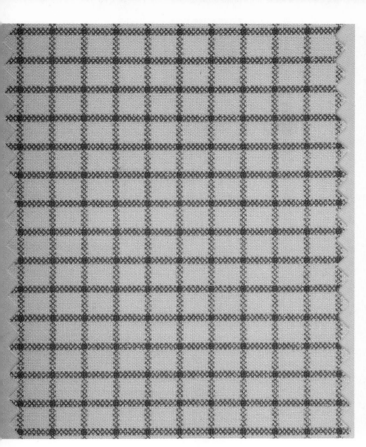

Cotton. USA. 1965.

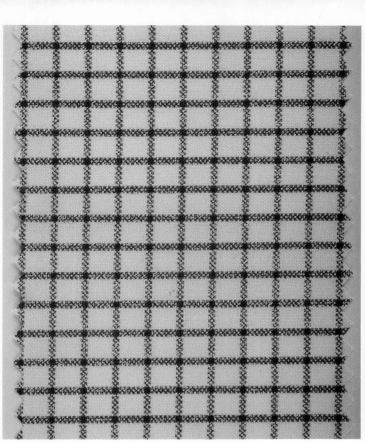

Cotton. USA. 1965.

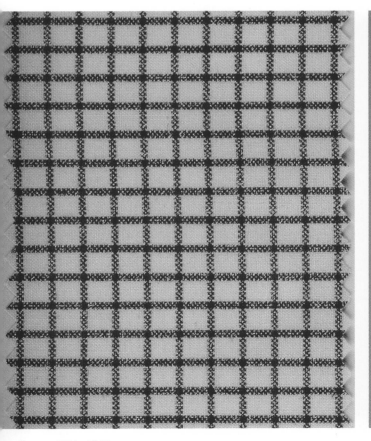

Cotton. USA. 1965.

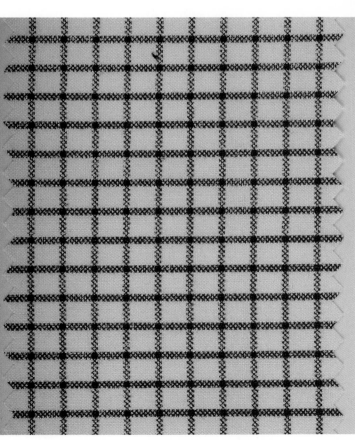

Cotton. USA. 1965.

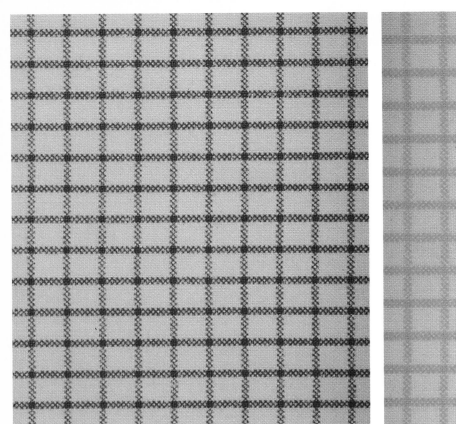

Cotton. USA. 1965.

Cotton. USA. 1965.

Cotton. USA. 1965.

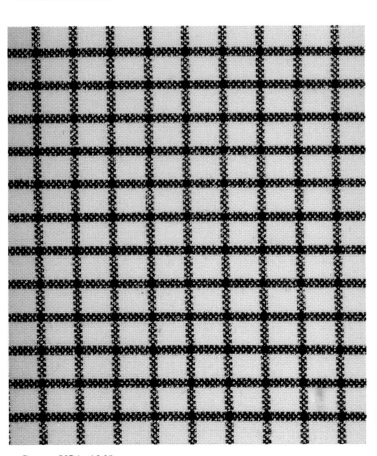

Cotton. USA. 1965.

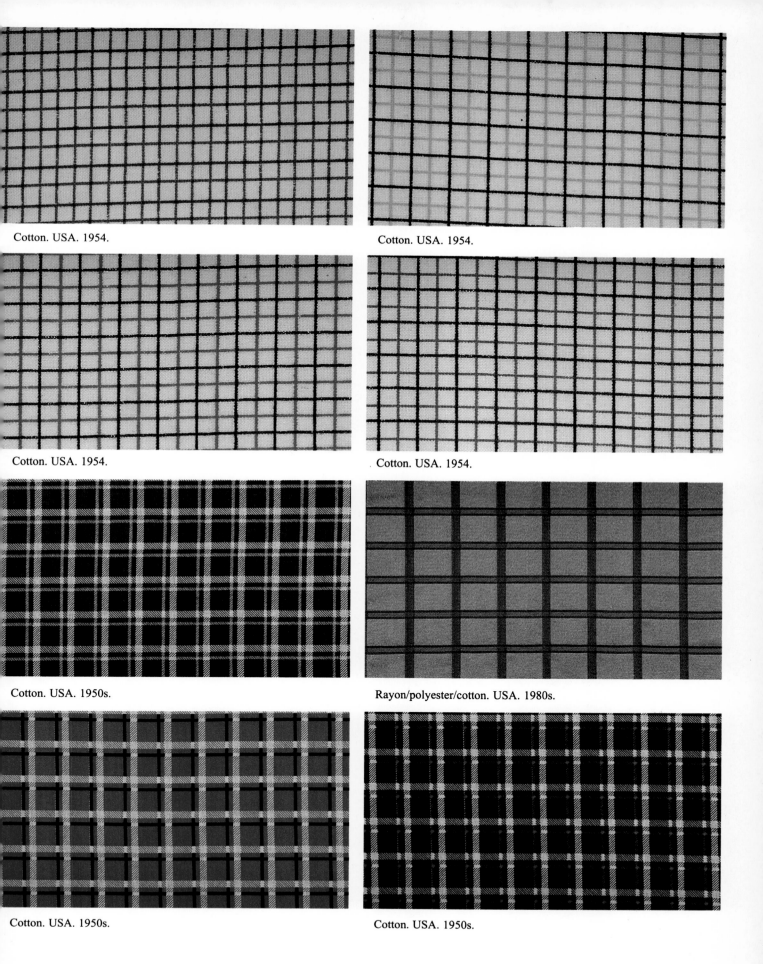

Cotton. USA. 1954.

Cotton. USA. 1954.

Cotton. USA. 1954.

Cotton. USA. 1954.

Cotton. USA. 1950s.

Rayon/polyester/cotton. USA. 1980s.

Cotton. USA. 1950s.

Cotton. USA. 1950s.

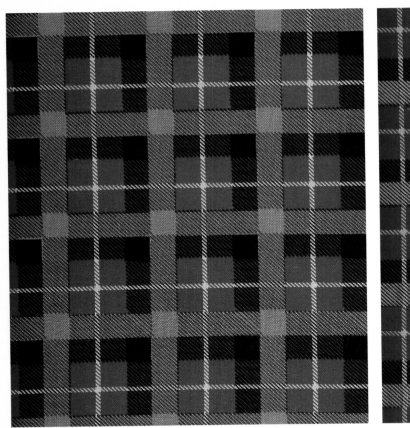

Cotton. USA. 1950s.

Cotton. USA. 1950s.

Cotton. USA. 1950s.

Cotton. USA 1935.

18

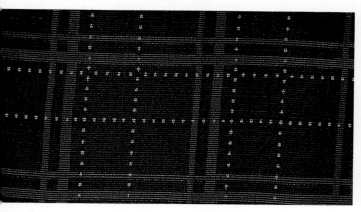

Silk/polyester. USA. 1950s.

Silk/polyester. USA. 1950s.

Polyester blend. USA. 1950s.

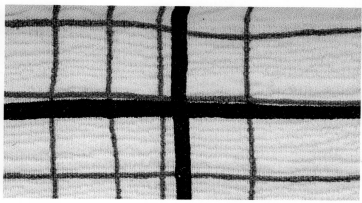

Cotton blend. France. 1967.

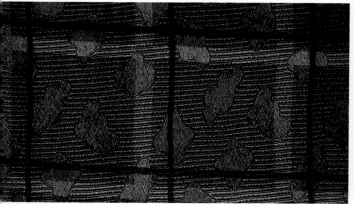

Silk. USA. 1950s.

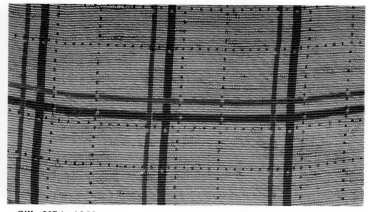

Silk. USA. 1950s.

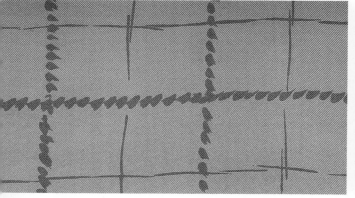

Cotton. Italy. 1964.

Silk. France. 1967.

gingham, squares, and checks

Cotton. USA. 1950s.

Silk. USA. 1950s.

Gingham

Cotton/polyester. France. 1959.

Wool blend. France. 1963.

Cotton. France. 1950s.

Silk/polyester. France. 1963.

21

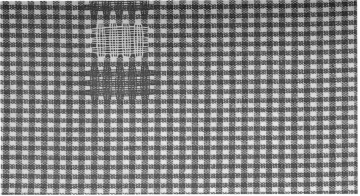

Cotton/rayon blend. USA. 1959.

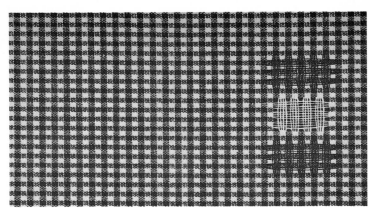

Cotton/rayon blend. USA. 1959.

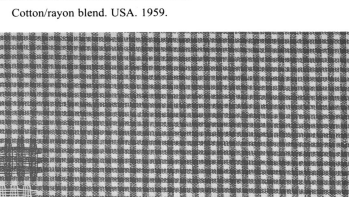

Cotton/rayon blend. USA. 1959.

Cotton/rayon blend. USA. 1959.

Cotton. USA. 1965.

Cotton. USA. 1965.

Cotton. USA. 1965.

Cotton. USA. 1965.

Squares

Cotton. USA. 1950s.

Cotton. USA. 1950s.

Cotton. USA. 1950s.

Cotton. Italy 1965.

Rayon. USA. 1980s.

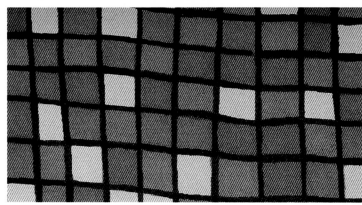

Silk. France. 1963.

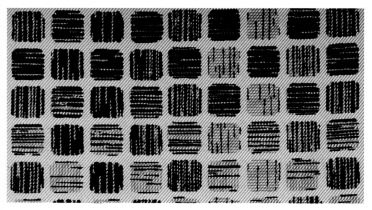

Rayon/silk blend. France. 1966.

Cotton. France. 1963.

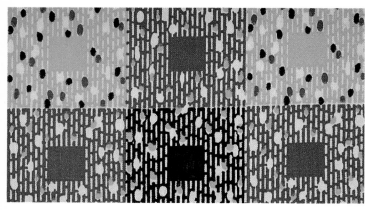

Cotton. USA. 1954.

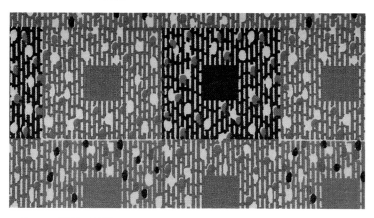

Cotton. USA. 1954.

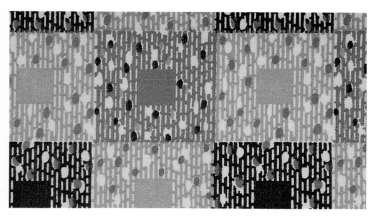

Cotton. USA. 1954.

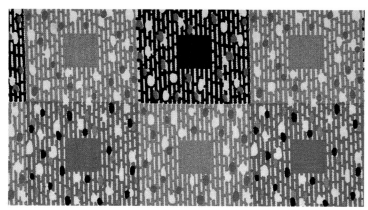

Cotton. USA. 1954.

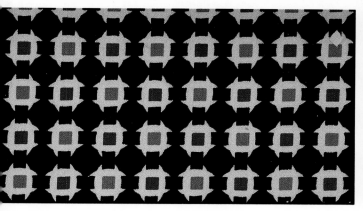

Cotton. USA. 1950s.

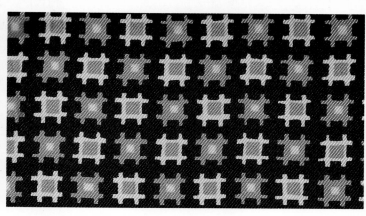

Rayon/silk blend. France. 1966.

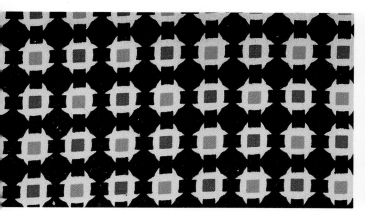

Cotton. USA. 1950s.

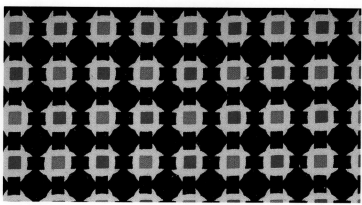

Cotton. USA. 1950s.

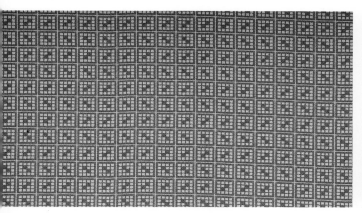

Cotton/rayon blend. USA. 1954.

Cotton. France. 1920s.

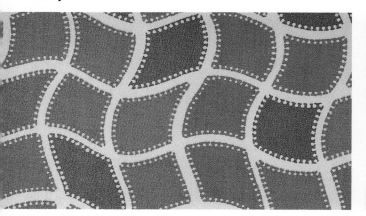

Cotton. USA. 1948.

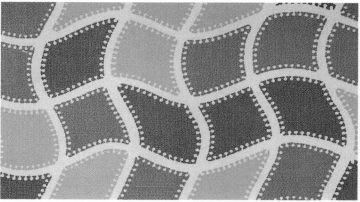

Cotton. USA. 1948.

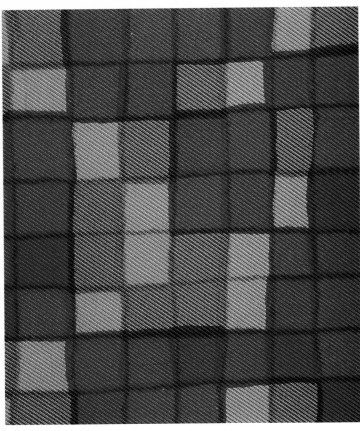

Rayon/silk blend. France. 1966.

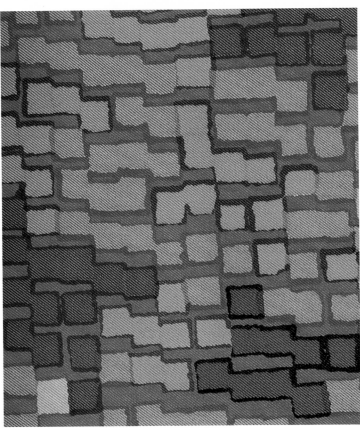

Silk. Italy. 1967.

Silk. Italy. 1967.

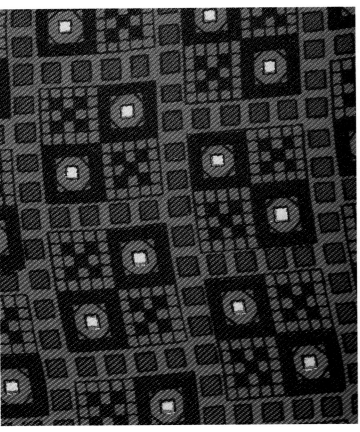

Rayon/silk blend. France. 1966.

Checks

Cotton. USA. 1954.

Cotton. USA. 1954.

Cotton. USA. 1954.

Cotton. USA. 1954.

Cotton/rayon blend. USA. 1950s.

Cotton/rayon blend. USA. 1950s.

Cotton/rayon blend. USA. 1950s.
Cotton/rayon blend. USA. 1950s.

Cotton/rayon blend. USA. 1950s.
Cotton/rayon blend. USA. 1950s.

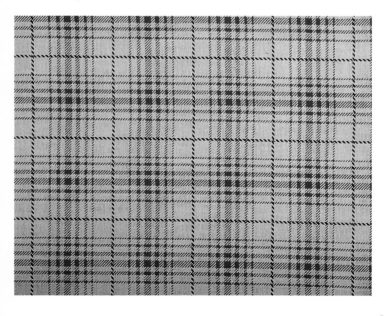

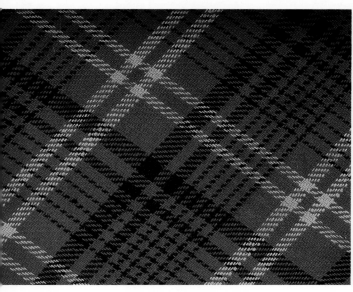

Cotton/polyester blend. USA. 1960s.

Cotton. Italy. 1968.

Rayon/cotton. USA. 1980s.
Rayon/cotton. USA. 1980s.

Cotton/polyester. USA. 1980s.
Rayon/cotton/polyester. USA. 1980s.

Cotton. USA. 1935.

Cotton. USA. 1935.

Cotton. USA. 1935.

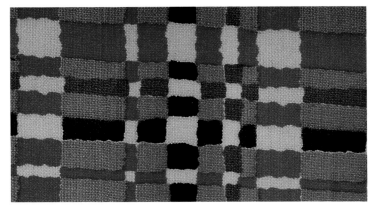

Cotton. USA. 1935.

Cotton. USA. 1950s.

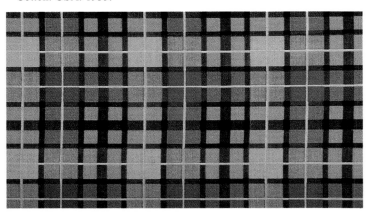

Cotton. USA. 1950s.

Cotton. USA. 1950s.

Cotton. USA. 1950s.

harlequin/diamonds

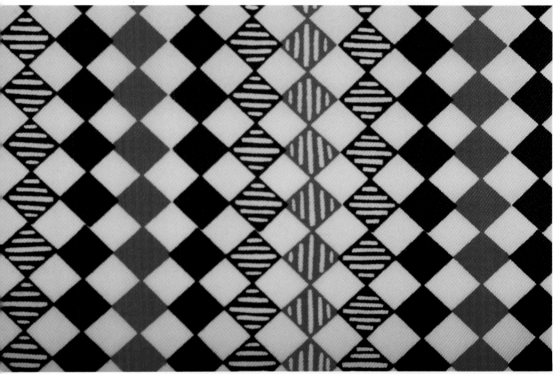

Cotton/rayon blend. USA. 1954.

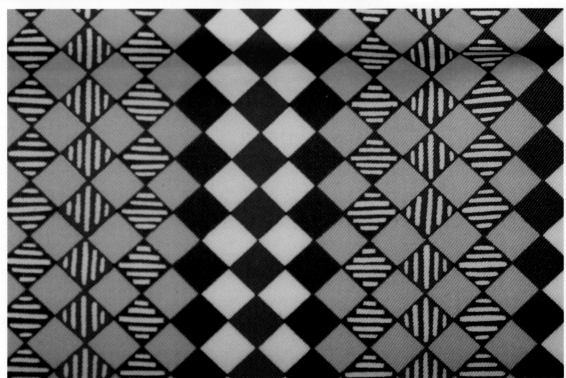

Cotton/rayon blend. USA. 1954.

Cotton/rayon blend. USA. 1954.

Cotton/rayon blend. USA. 1954.

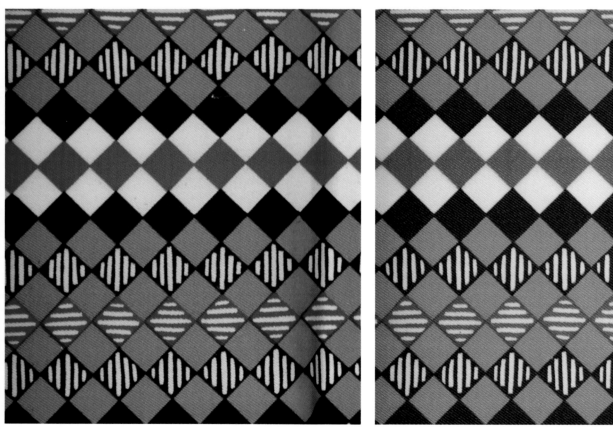

Cotton/rayon blend. USA. 1954.

Cotton/rayon blend. USA. 1954.

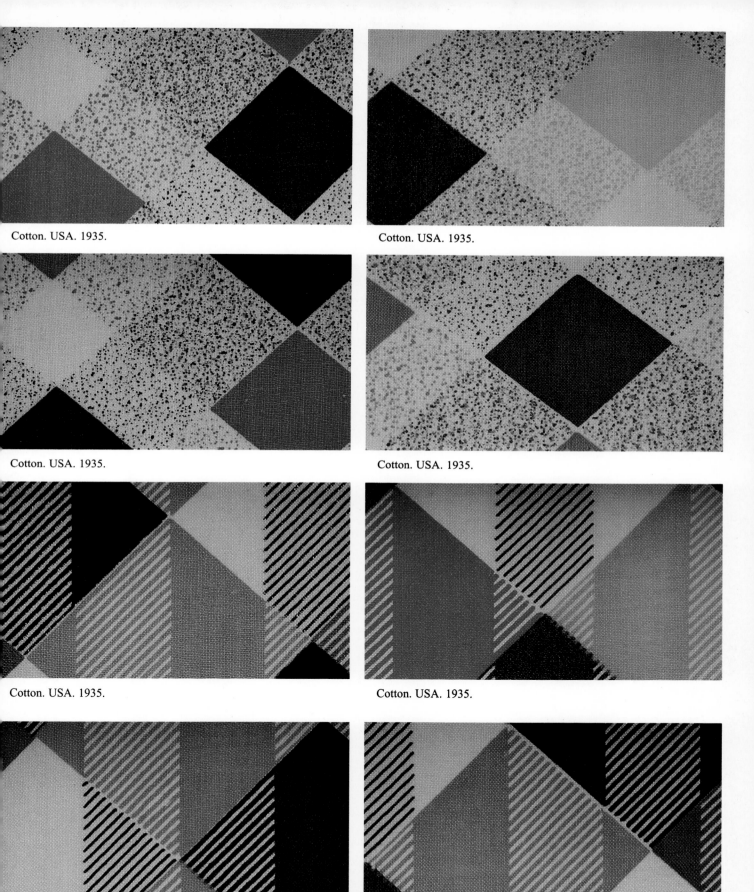

Cotton. USA. 1935.

Cotton. USA. 1935.

Cotton. USA. 1935.

Cotton. USA. 1935.

Cotton. USA. 1935.

Cotton. USA. 1935.

Cotton. USA. 1935.

Cotton. USA. 1935.

Cotton. USA. 1954.

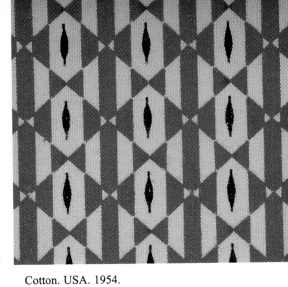

Cotton. USA. 1954.

Cotton. USA. 1954.
Cotton. USA. 1935.

Cotton. USA. 1954.
Cotton. USA. 1935.

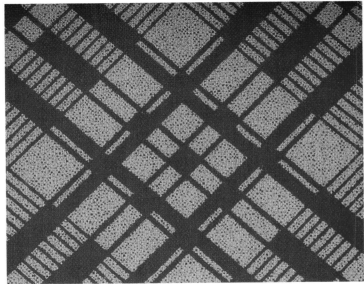

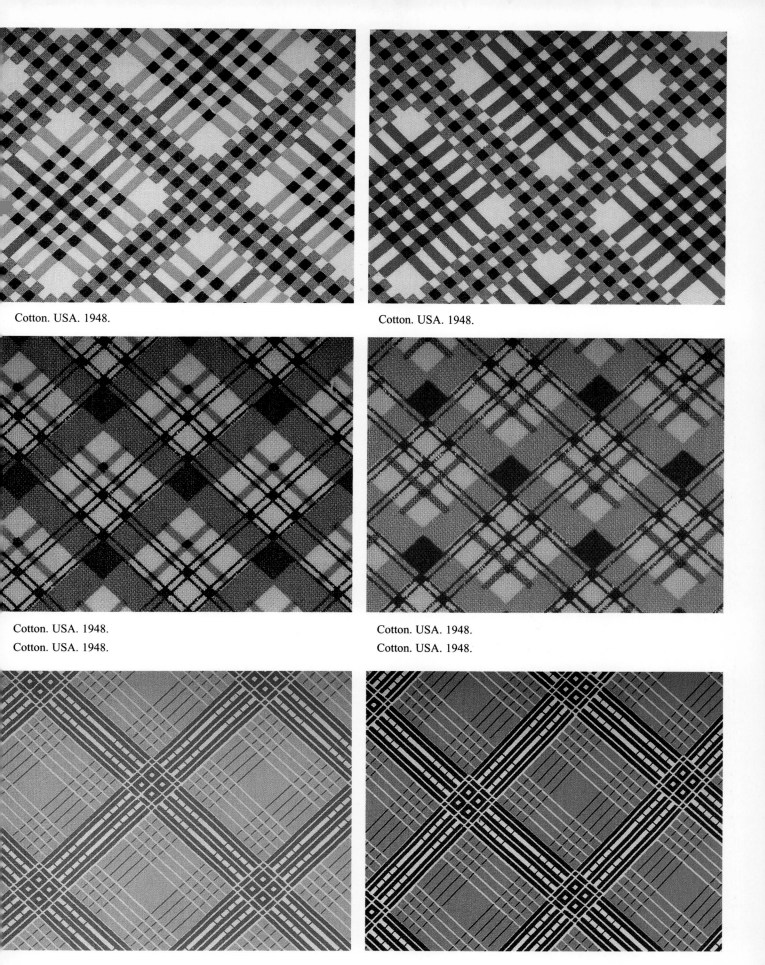

Cotton. USA. 1948.

Cotton. USA. 1948.

Cotton. USA. 1948.
Cotton. USA. 1948.

Cotton. USA. 1948.
Cotton. USA. 1948.

Cotton/rayon blend. USA. 1959.

Cotton. USA. 1948.

Cotton. USA. 1948.

Cotton. USA. 1950s.

Cotton. USA. 1935.

Cotton. USA. 1935.

Cotton. USA. 1935.

Cotton. USA. 1935.

Cotton. USA. 1948.

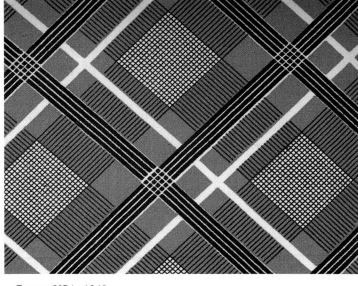

Cotton. USA. 1948.

Linen/rayon. France. 1966.

Cotton. USA. 1963.

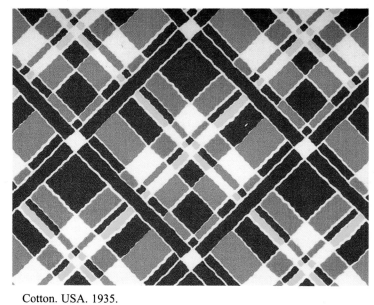

Cotton. USA. 1935.

Cotton. USA. 1963.

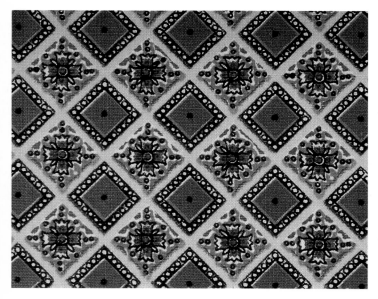

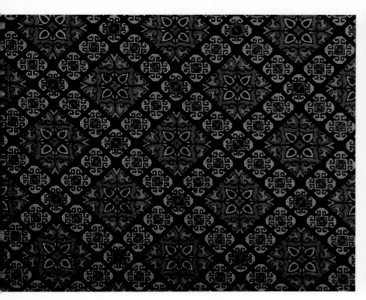

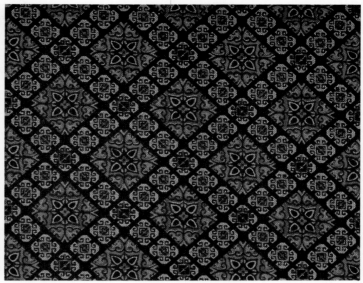

Cotton. USA. 1963.

Cotton. USA. 1963.

Polyester. France. 1966.

Cotton/rayon blend. USA. 1950s.

Cotton/rayon blend. USA. 1950s.

Cotton/rayon blend. USA. 1950s.

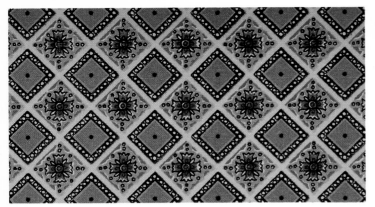

Cotton. USA. 1963.

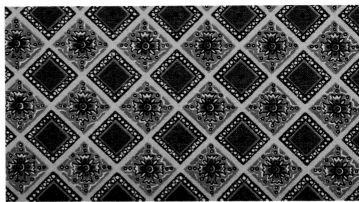

Cotton. USA. 1963.

Cotton. USA. 1935.

Cotton. France. 1920s.

Silk/rayon blend. France. 1966.

Polyester. France. 1966.

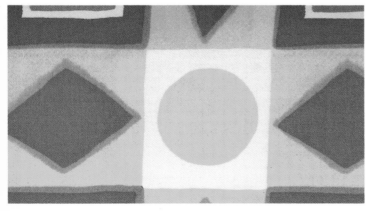

Cotton. France. 1963.

Cotton. France. 1963.

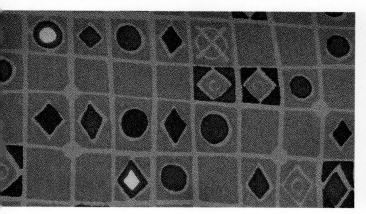

Silk. France. 1967.

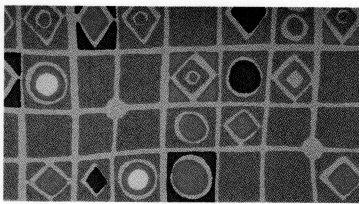

Silk. France. 1966.

Rayon/silk blend. France. 1966.

Silk. France. 1966.

Wool. France. 1966.

Silk. France. 1967.

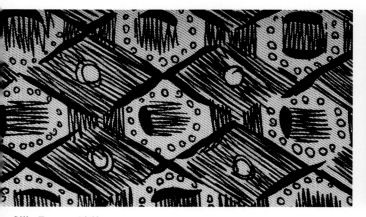

Silk. France. 1963.

Polyester. France. 1966.

Silk/rayon blend. France. 1966.

Cotton. France. 1963.

Felt. Italy. 1968.

Manmade fibers. Italy. 1967.

Silk. France. 1967.

Silk. France. 1967.

Silk/polyester blend. France. 1966.

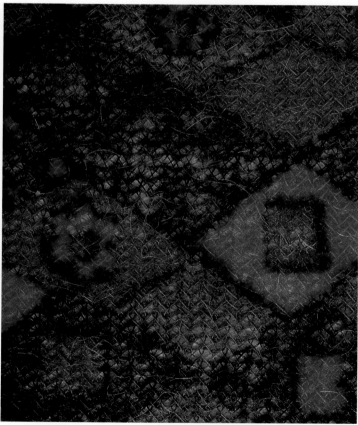

Mohair/Wool. France. 1966.

43

hound's-tooth

Silk. France. 1964.

Cotton/rayon. France. 1963.

Silk. France. 1967.

Polyester. France. 1966.

Silk. France. 1967.

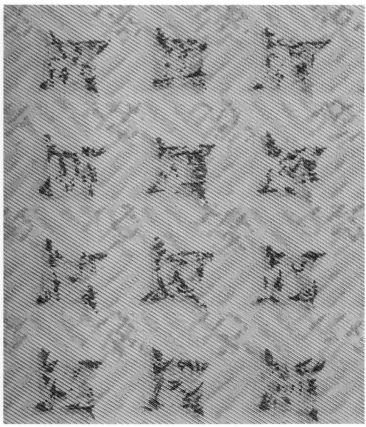

Silk. France. 1967.

Silk. France. 1967.

Silk/cotton. France. 1964.

Cotton. France. 1963.

Silk blend. France. 1966.

novelty and floral

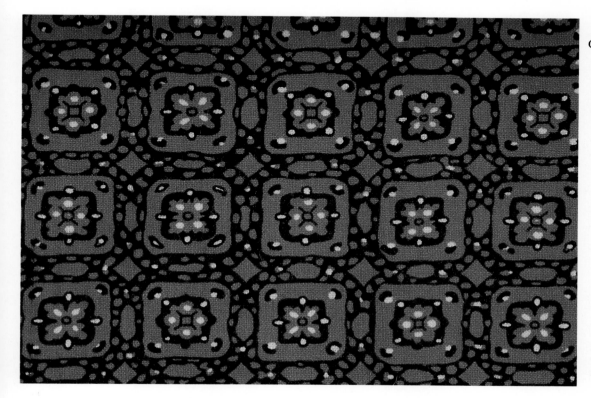

Cotton. USA. 1950s.

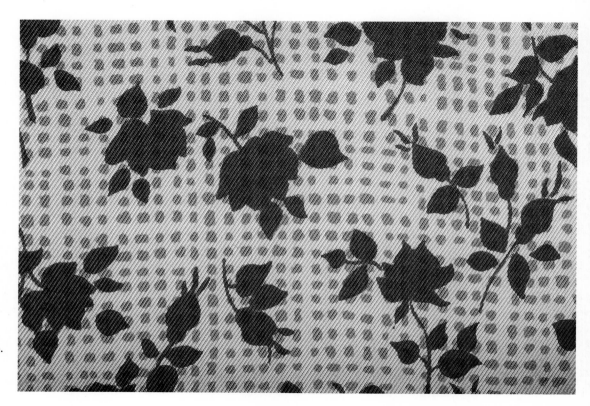

Polyester. France. 1966.

Novelty

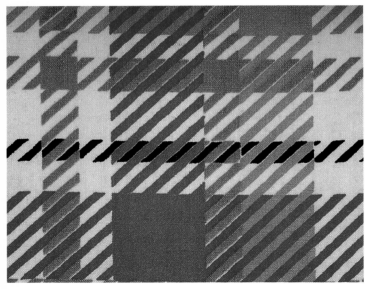

Cotton. USA. 1935.

Cotton. USA. 1935.
Cotton. USA. 1935.

Cotton. USA. 1935.

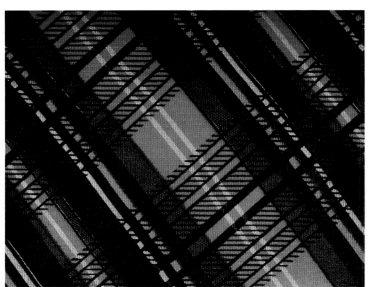

Cotton. USA. 1920s.

Cotton/rayon/linen. USA. 1980s.

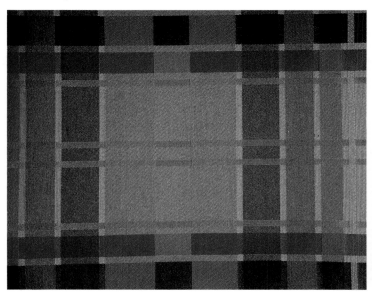

Silk/rayon blend. France. 1966.

Wool blend. France. 1963.

Cotton. Italy. 1965.

Cotton. France. 1962.

Polyester blend. France. 1966.

Cotton. France. 1964.

Dacron/cotton. USA. 1980s.

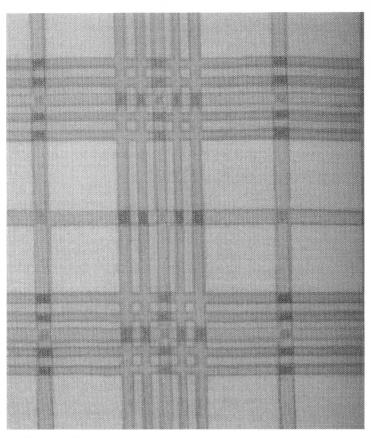

Silk. Italy. 1965.

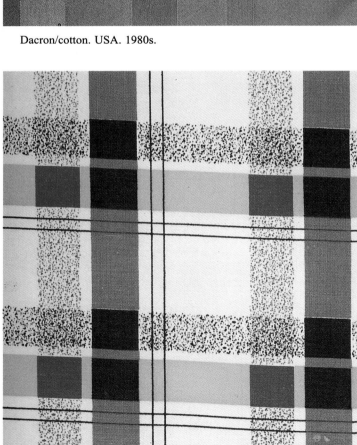

Cotton. USA. 1935.

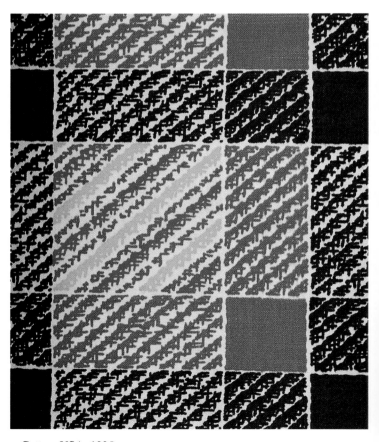

Cotton. USA. 1935.

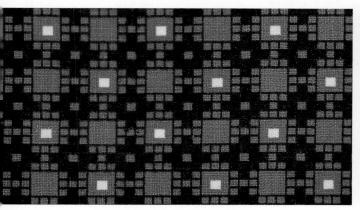
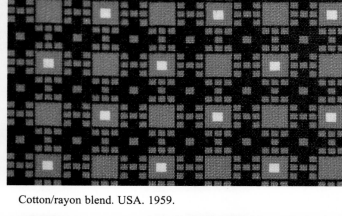

Cotton/rayon blend. USA. 1959.

Cotton/rayon blend. USA. 1959.

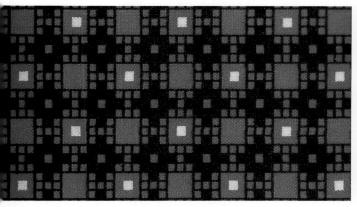
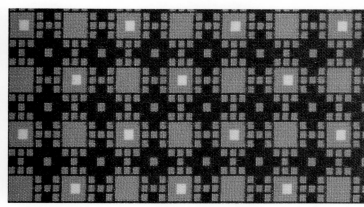

Cotton/rayon blend. USA. 1959.

Cotton/rayon blend. USA. 1959.

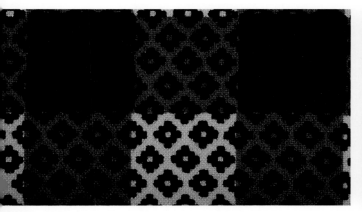

Polished cotton. USA. 1950s.

Polished cotton. USA. 1950s.

Polished cotton. USA. 1950s.

Polished cotton. USA. 1950s.

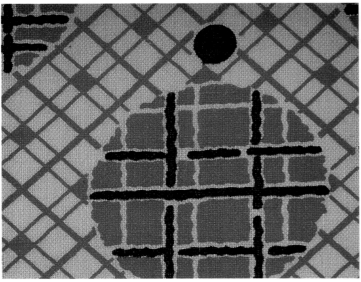

Cotton. USA. 1935.

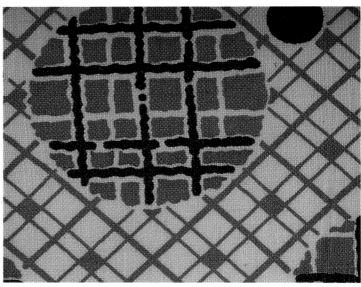

Cotton. USA. 1935.

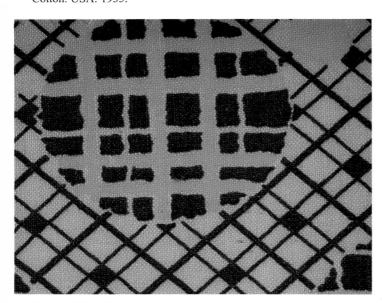

Cotton. USA. 1935.

Cotton. USA. 1935.

Cotton. USA. 1935.

Cotton. USA. 1935.

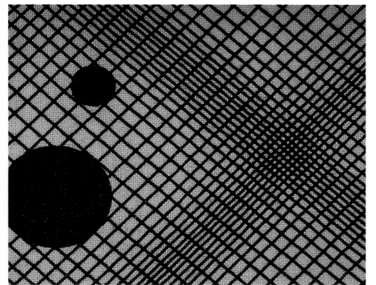

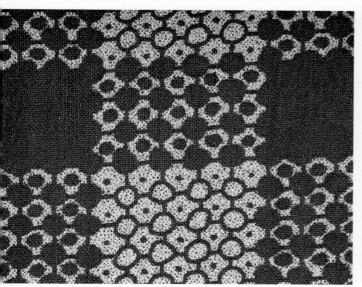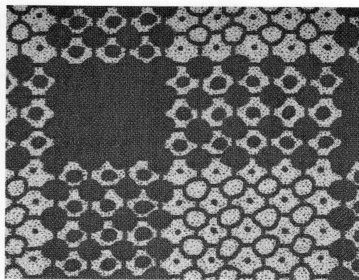

Cotton. USA. 1935.　　　　　　　　　　　　　Cotton. USA. 1935.

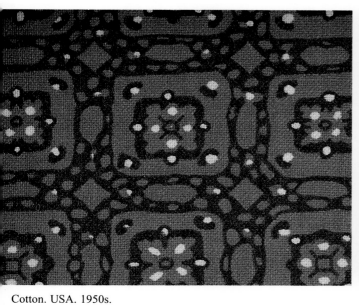

Cotton. USA. 1950s.　　　　　　　　　　　　Cotton. USA. 1950s.

Cotton. USA. 1950s.　　　　　　　　　　　　Cotton. USA. 1950s.

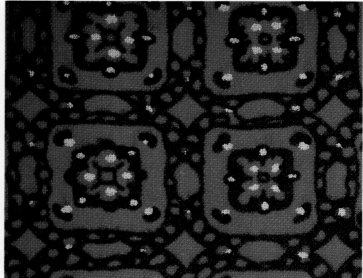

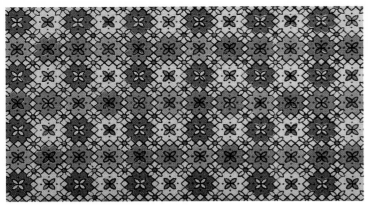

Cotton. USA. 1965.

Cotton. USA. 1965.

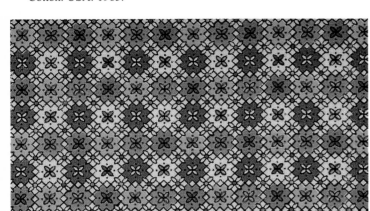

Cotton. USA. 1965.

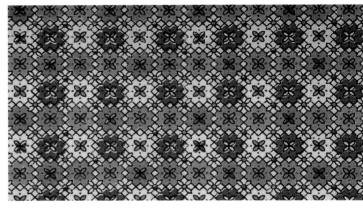

Cotton. USA. 1965.

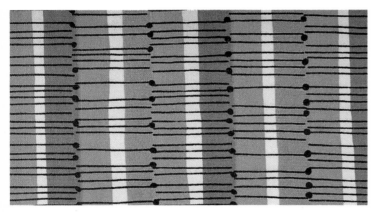

Cotton. USA. 1950s.

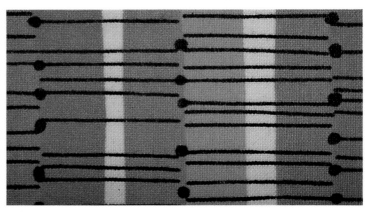

Cotton. USA. 1950s.

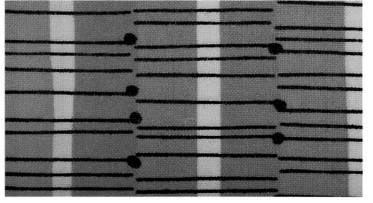

Cotton. USA. 1950s.

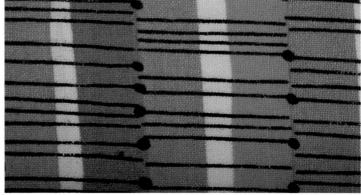

Cotton. USA. 1950s.

Rayon. France. 1966.

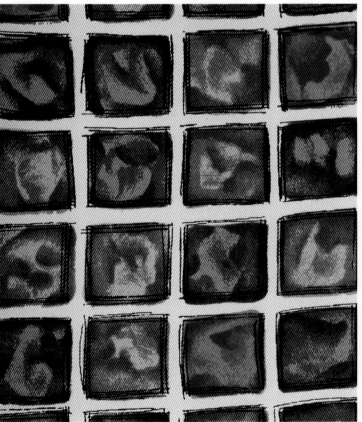

Polyester. France. 1966.

Cotton. France. 1961.

Silk. France. 1963.

Cotton. France. 1963.

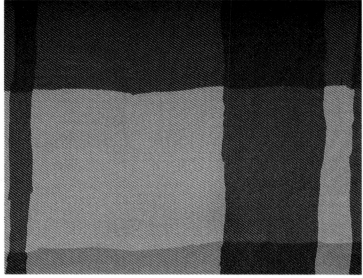

Silk. France. 1967.

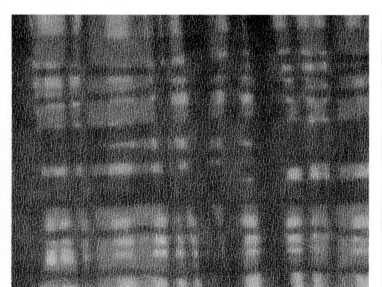

Wool. France. 1966.
Cotton/polyester. France. 1962.

Cotton. France. 1961.
Cotton/polyester. France. 1962.

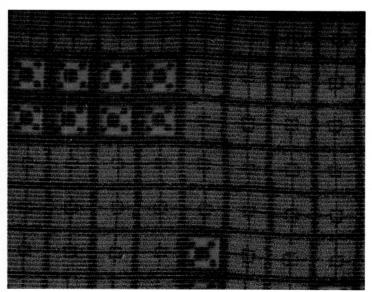

Silk/polyester blend. France. 1966.

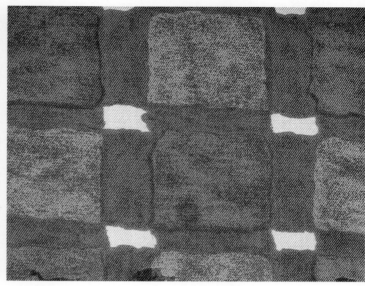

Silk. France. 1959.

Nylon. France. 1963.
Wool blend. France 1966

Silk. France. 1962.
Silk. France. 1963.

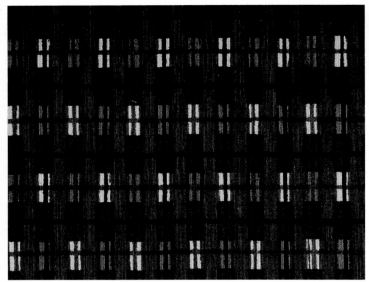

Cotton. France. 1962.

Silk/polyester. France. 1959.

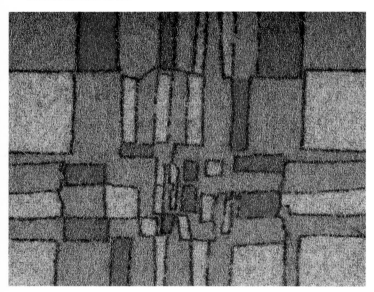

Wool France 1966

Polyester. France. 1966.

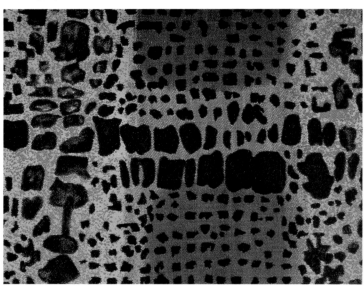

Cotton blend. France. 1963..

Silk. France. 1967.

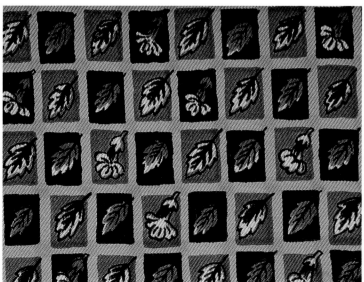

Cotton. USA. 1950s.

Cotton. France. 1963.

Cotton/polyester. France. 1966.

Silk. France. 1962.

Silk blend. Italy. 1968.

Polyester. France. 1966.

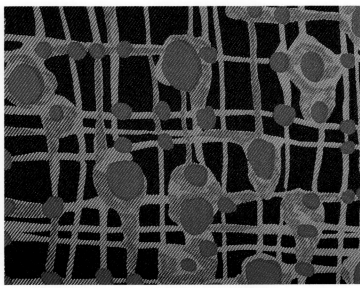

Cotton/silk. France. 1962.

Cotton/polyester. France. 1962.

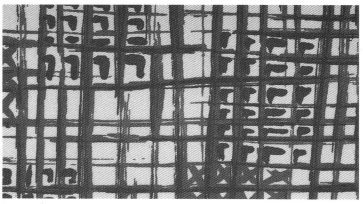

Silk. France. 1963.

Cotton. France. 1962.

Silk. France. 1959.

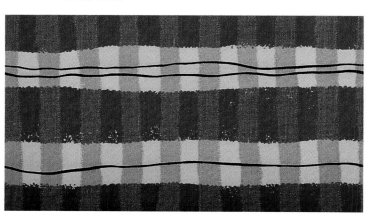

Cotton. USA. 1950s.

Cotton/polyester. France. 1964.

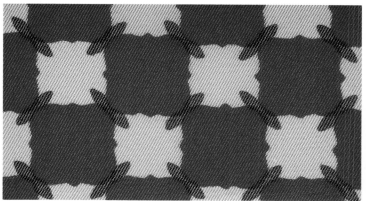

France. 1967.

Cotton. France. 1964.

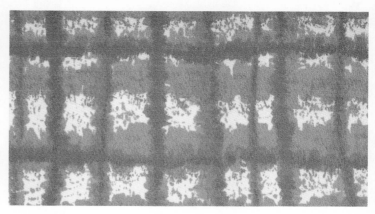

Cotton. France. 1962.

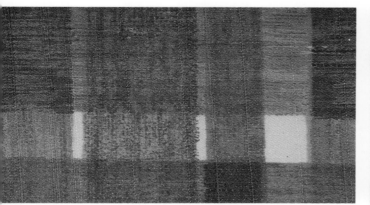

Cotton. France. 1962.

Silk. Italy. 1967.

Cotton. France. 1963.

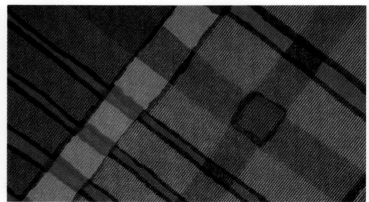

Silk. France. 1967.

Cotton. France. 1950s.

Cotton/polyester. France. 1963.

Nylon. France. 1962.

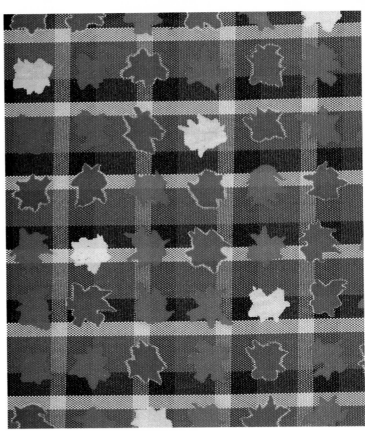

Rayon. France. 1961.

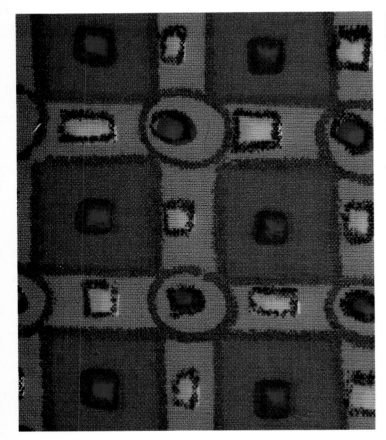

Cotton. France. 1964.

Silk/polyester blend. France. 1966.

Cotton/polyester. France. 1966.

Cotton/polyester. France. 1966.

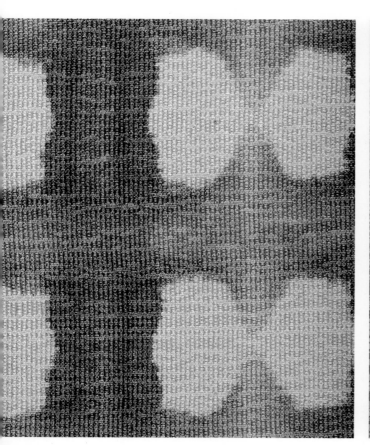

Cotton/acrylic. France. 1962.

Cotton. France. 1950s.

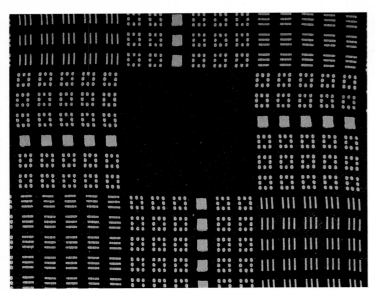

Silk. USA. 1960s.

Silk. Italy. 1965.

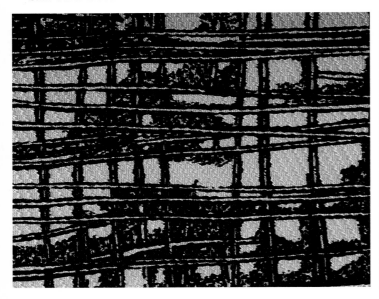

Polyester. France. 1966.

Silk. USA. 1960s.

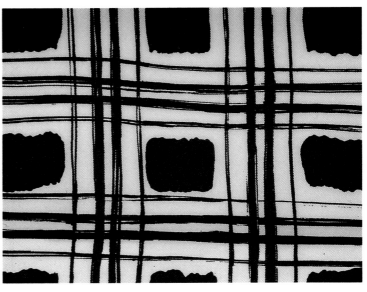

Silk. USA. 1960s.

Silk. USA. 1960s.

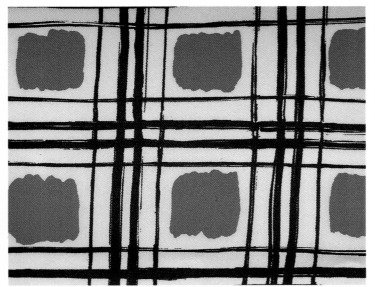

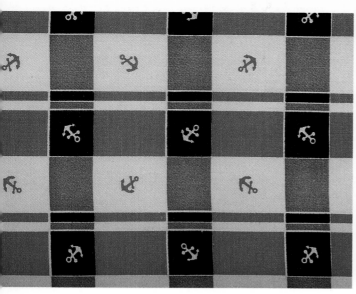

Cotton. USA. 1935.

Cotton/rayon. France. 1959.

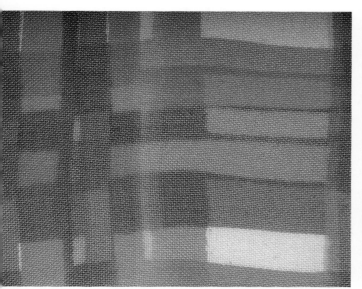

Cotton/nylon blend. France. 1966.

Polyester. France. 1966.

Silk. France. 1963.

Cotton. France. 1961.

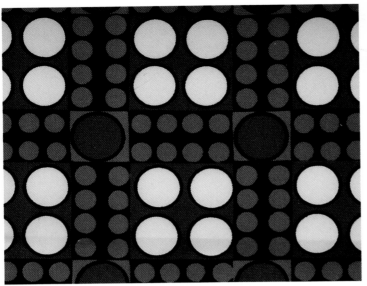

Silk. Italy. 1960s.

Silk. Italy. 1960s.

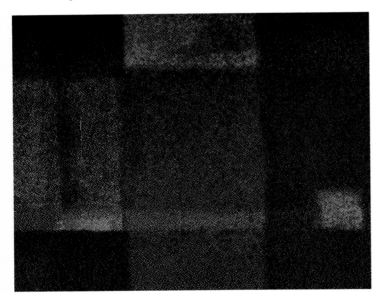

Silk/polyester blend. France. 1966.

Cotton. France. 1966.

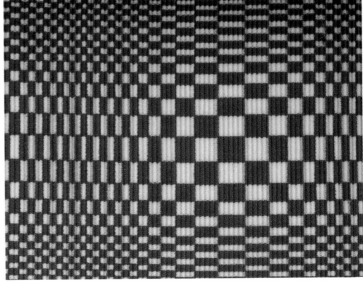

Cotton. France. 1962.

Cotton. France. 1966.

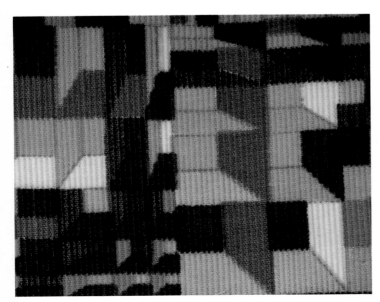

France. 1967.

Silk. France. 1963.

Cotton. USA. 1935.
Cotton. France. 1964.

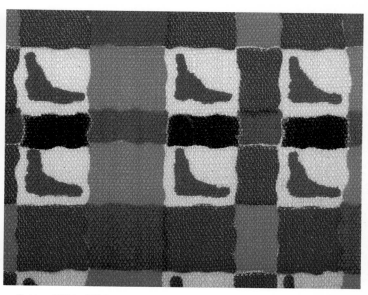

Cotton. USA. 1935.
Silk. France. 1967.

Cotton. USA. 1935.

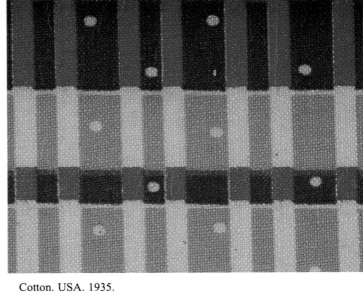

Cotton. USA. 1935.

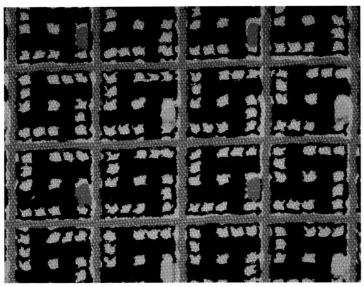

Cotton. USA. 1935.
Cotton. USA. 1948.

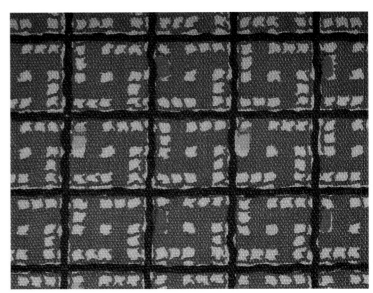

Cotton. USA. 1935.
Cotton. USA. 1948.

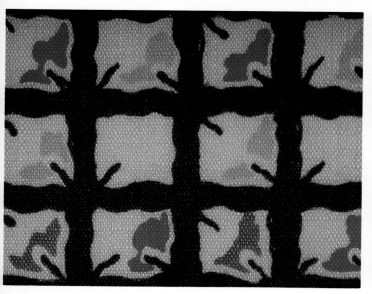

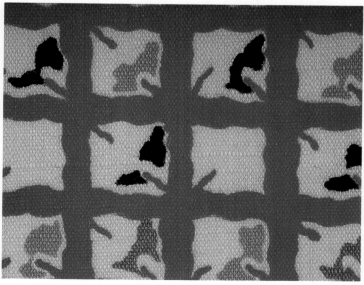

Cotton. USA. 1948.

Cotton. USA. 1948.

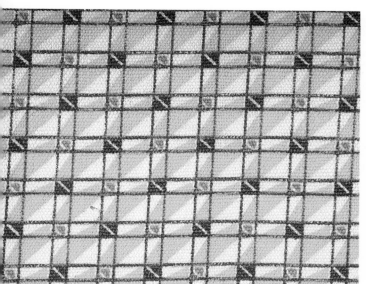

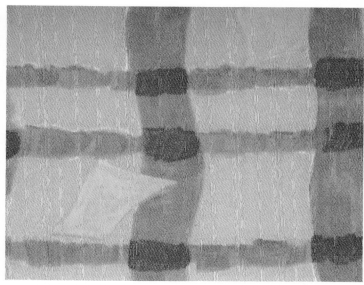

Cotton. USA. 1935.

Cotton. France. 1962.

Cotton/rayon blend. USA. 1959.

Cotton. USA. 1980s.

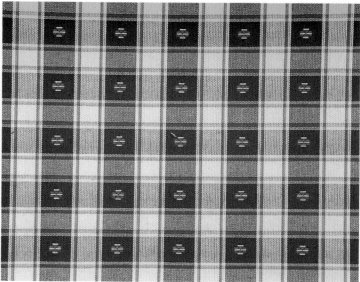

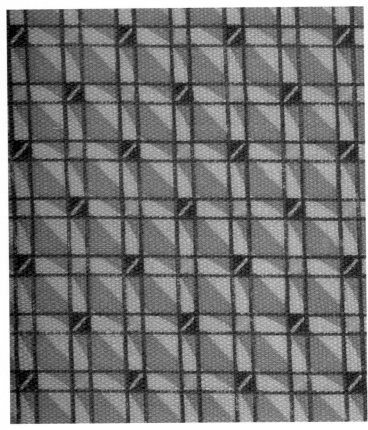

Cotton. USA. 1935.

Silk. France. 1963.

Cotton/polyester. France. 1962.

Cotton/rayon. France. 1962.

Floral

Cotton. USA. 1935.

Cotton. USA. 1935.

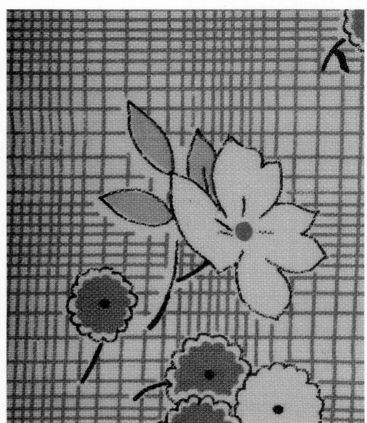

Cotton. USA. 1935.

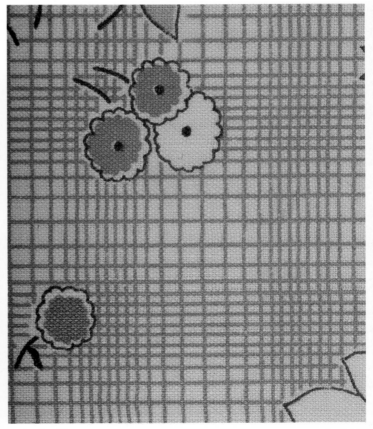

Cotton. USA. 1935.

Cotton/rayon. USA. 1959.

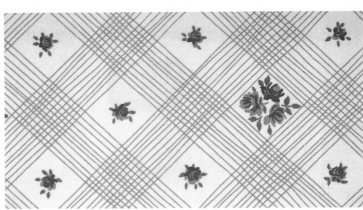

Cotton/rayon. USA. 1959.

Cotton/rayon. USA. 1959.

Cotton/rayon. USA. 1959.

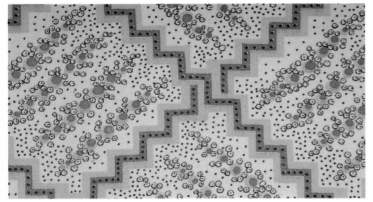

Cotton. USA. 1954.

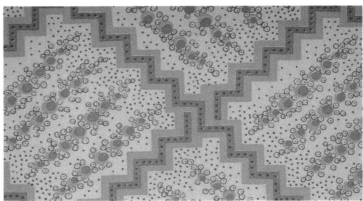

Cotton. USA. 1954.

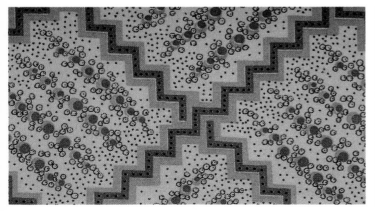

Cotton. USA. 1954.

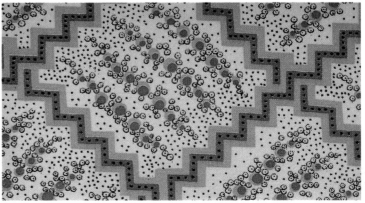

Cotton. USA. 1954.

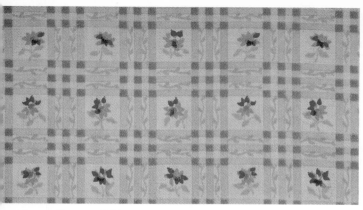

Cotton. France. 1964.

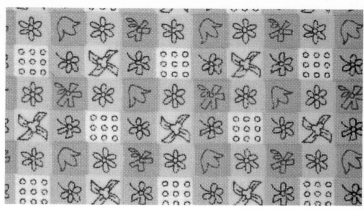

Cotton. USA. 1954.

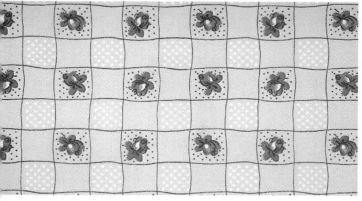

Cotton/rayon blend. USA. 1959.

Cotton. USA. 1954.

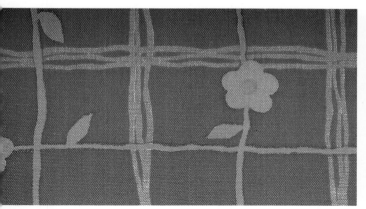

Silk. Italy. 1965.

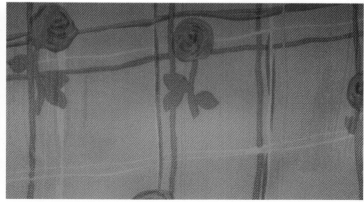

Silk. Italy. 1964.

Cotton/rayon blend. USA. 1950s.

Cotton/rayon blend. USA. 1950s.

Cotton. USA. 1935.

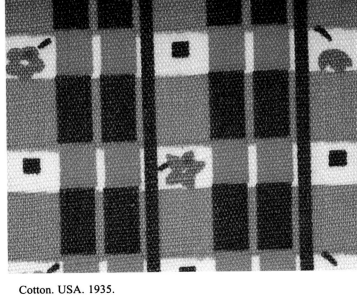

Cotton. USA. 1935.

Cotton. France. 1962.
Cotton/rayon blend. USA. 1950s.

Cotton blend. France. 1964.
Cotton. France. 1966.

Cotton. France. 1966.

Cotton. France. 1964.

Silk. France. 1963.
Cotton/polyester. France. 1950s.

Cotton. France. 1966.
Cotton. France. 1963.

Cotton. Italy. 1963.

Silk. France. 1963.

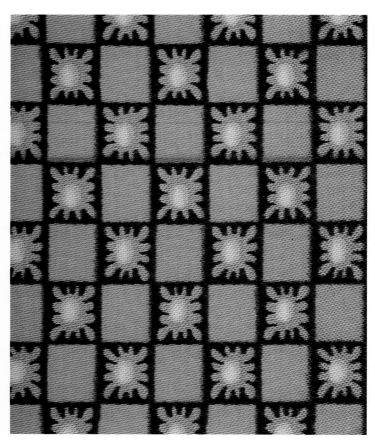

Nylon/cotton. France. 1964.

Silk. France. 1967.

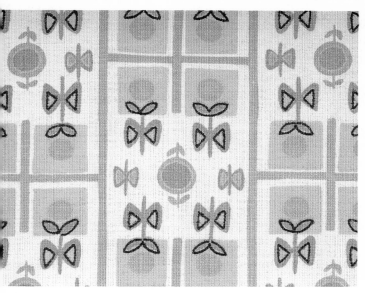

Cotton/polyester. France. 1963.

Cotton. France. 1963.

Cotton/rayon blend. 1950s.

Cotton. France. 1963.

Silk. Italy. 1965.

Silk. Italy. 1965.

77

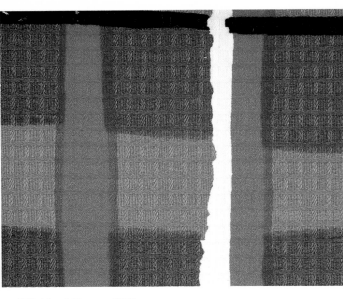

Silk blend. France. 1962.

Silk. Italy. 1965.

Silk. Italy. 1967.
Polyester. France. 1966.

Cotton. France. 1920s.
Cotton. France. 1963.

scottish tartans

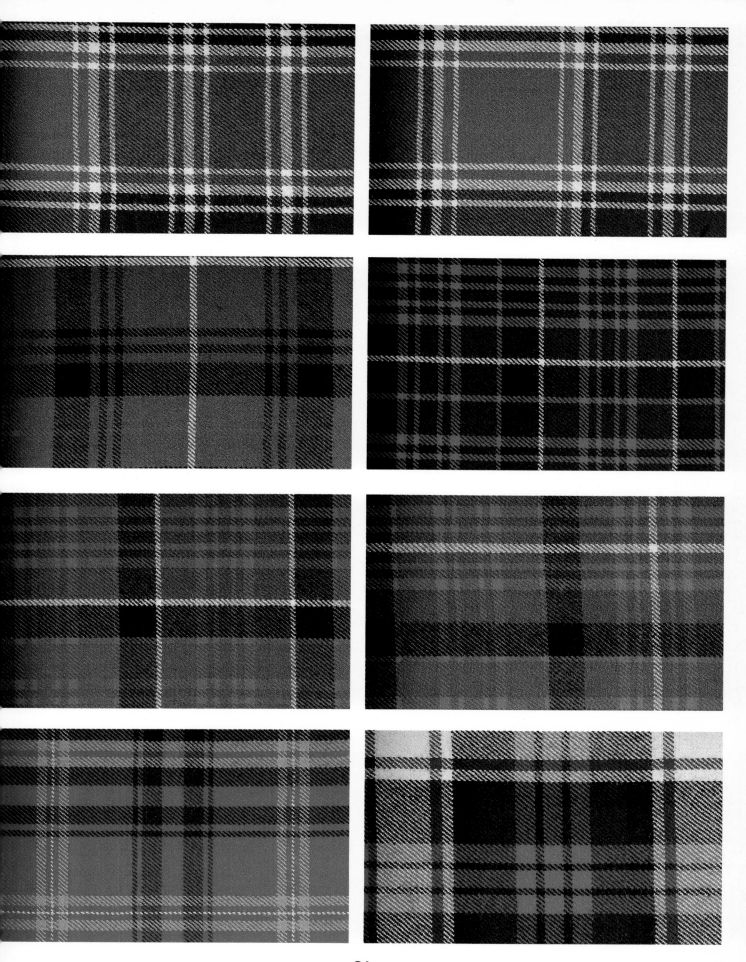

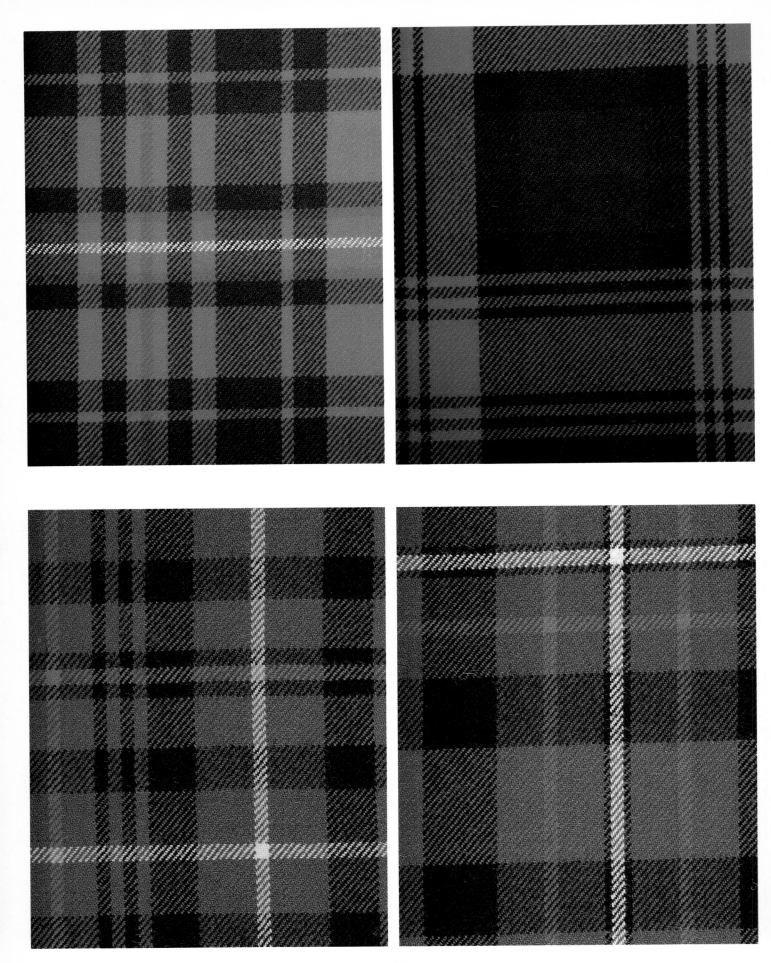

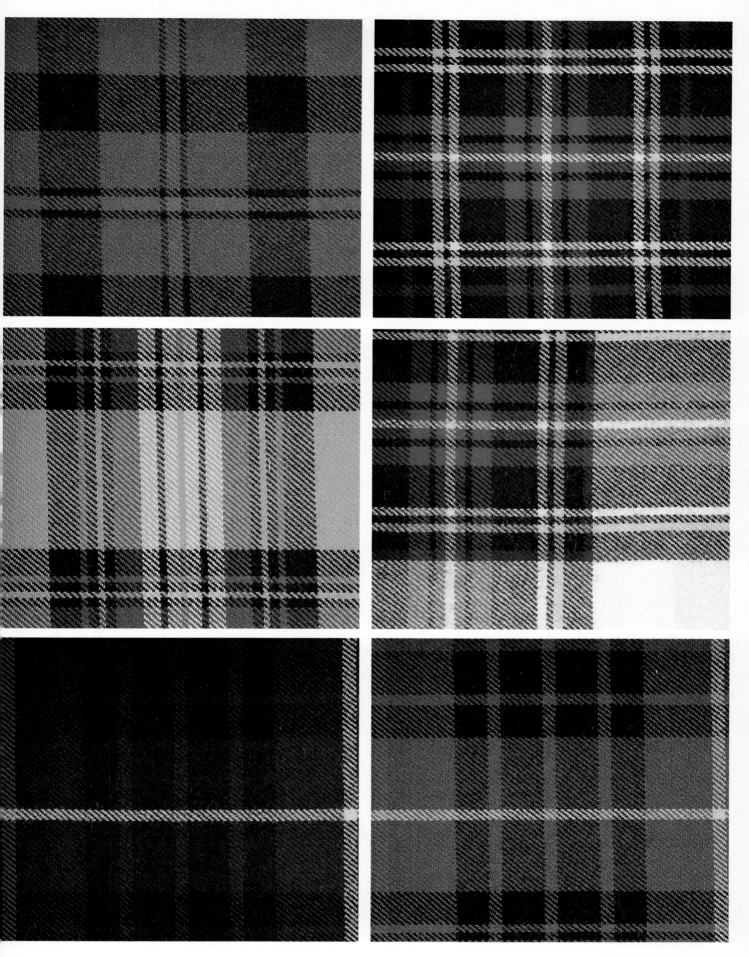